SPACE DOGS

The story of the celebrated
canine cosmonauts

Laurence King Publishing

Published in 2019 by
Laurence King Publishing Ltd
361–373 City Road
London EC1V 1LR

Tel +44 20 7841 6900
Fax +44 20 7841 6910

enquiries@laurenceking.com
www.laurenceking.com

Photographs © Martin Parr 2019
With thanks to Louis Little
Text © Laurence King Publishing

Text by Richard Hollingham

ISBN 978 1 78627 411 3

Commissioning editor: Zara Larcombe
Senior editor: Gaynor Sermon
Consultant editor: Iryna Rybinkina
Design: Nicolas Franck Pauly
Production manager: Sian Smith

Printed in China

Foreword

September 2018, Martin Parr Foundation, UK

From the first moment I saw a piece of space dog ephemera I was hooked. It's such a great story. Russia was winning the space race that was taking place against the backdrop of the Cold War. Stray dogs were being plucked off the streets of Moscow and trained for space missions. Quite how this process was undertaken has always been a mystery to me.

Laika was the first dog to orbit the Earth, but she sadly perished on re-entry into the Earth's atmosphere. This fact was somewhat obscured by the Russian space agency for decades as they did not want the mission to be viewed as a failure. But Laika's demise did not hinder the advent of the space dog memorabilia industry. Russia needed heroes, and Laika – and later Belka and Strelka – fulfilled this role perfectly.

The amount and variety of space dog ephemera that was produced is mind boggling. Clocks, cigarette cases, ornaments and myriad dog-related items helped to validate Russia as the superior power in the Cold War.

If the response to the supposed triumph of Laika was extraordinary, things really ramped up to a whole new level with Belka and Strelka. Not only did these dogs survive the ordeal of orbiting space, but they could also be paraded and photographed upon their return, and thus became the new national heroes.

Russian society was not accustomed to the heroes and superstars that we are used to seeing in the West, and the space dogs easily fitted into this new role. Perhaps a useful way to really understand the impact that they had on Russian society is to draw a parallel with The Beatles or Mickey Mouse, those Western icons that generated huge quantities of memorabilia.

After Belka and Strelka, other dogs went into space but the momentum slowed, yet this notorious pair of stray dogs became icons of the whole space dogs programme.

So how did I come across all of the material in my collection? Apart from picking up some items of space dog memorabilia in the famous Moscow flea markets, my main source has been the internet, and over the past 20 years regular searches have unearthed many items for the collection.

My quest was consolidated when I connected with Natalie, an e-bay seller who would always have extremely good space dog items on offer. These have been posted to me and even collected, via a third party, on one of my many visits to Moscow. These days I continue to trawl the internet, but most objects I own already, or they don't really excite me, so what you see in the following pages is the sum total of a 20-year obsession. I hope that some of the magic of the space dogs phenomenon will rub off on the reader.

Martin Parr

КИ В
ЮСЕ

Dogs in Space

In the 1950s the space race between the USA and the USSR was well and truly on, and was for both nations a matter of pride and propaganda. But before man ventured into the cosmos, his four-legged friends would pave the way for space exploration.

One Man and his Space Dog

12 April 1961, Baikonur Cosmodrome, Kazakhstan

Yuri Gagarin lay on the couch of his cramped Vostok space capsule, ready to make history. But the experience of preparing to launch into orbit was turning out to be, well ... a little bit boring.

In the control bunker, head of the Soviet rocket programme, Sergei Korolev, was growing increasingly frustrated by the delays. While they were waiting, one of the mission controllers asked Gagarin how he was faring. 'If there was some music,' replied the 27-year-old fighter pilot, 'I could stand it a little better.'

After a 20-minute musical interlude of Russian love songs, Gagarin heard a faint pop as the service towers retracted and fuel valves opened in the rocket below. Then the cosmonaut felt a gentle vibration as the launcher rose from the pad and a gradual acceleration as it headed into the clouds. 'Let's go!' he shouted, before the G-force made it difficult to speak.

Two-and-a-half minutes later, the protective casing covering the capsule fell away to reveal the planet below. It was the first time a human had seen the Earth from space with their own eyes. 'I see the clouds, the landing site,' Gagarin exclaimed. 'It's beautiful, what beauty!'

Although Gagarin was the first to describe the experience of leaving the Earth, he wasn't the first to make the journey into orbit. Nor was he the first to fly in the new

Vostok space capsule. Over the previous nine months, ten dogs had attempted the same journey. Only six had made it back alive.

These animals were among the final pioneers of a 20-year Soviet space dog programme. Most of the flights were secret and dozens of dogs lost their lives, but the knowledge gained from the missions still helps astronauts today. Before the giant leap for mankind, the small steps on both sides of the international space race were made by animals. And whereas the Russians used dogs, the Americans favoured primates.

A few weeks before Gagarin's flight, NASA scientists strapped a chimp called Ham (named after the Holloman Aerospace Medical Center) into an experimental Mercury capsule. They had trained Ham to pull two levers in response to flashing lights. If the chimp failed to pull the correct lever in time, he received an electric shock. After a 15-minute flight, the spacecraft splashed down in the ocean and scientists opened the hatch. Ham seemed to be smiling. Primate experts later described the look on his face as one of utter terror.

Although dogs couldn't be trained to operate levers, they proved much more amenable than chimps or monkeys to being strapped into rockets. Unlike Ham – whose mission was soon overshadowed by the flights of the silver-suited US Mercury astronauts – many space dogs also became international heroes. Even today, space dog Laika is as famous, maybe more so, than Gagarin or the first woman in space, Valentina Tereshkova.

Dogs with the Right Stuff

Summer 1950, Moscow suburbs

It was tough being one of the thousands of stray dogs on the streets of Moscow. Life was short and violent, with little food or shelter. Winter temperatures regularly dropped to below -20 degrees Celsius and hundreds of dogs died of cold or starvation.

In the post-war Soviet Union, few people could afford to keep pets and the number of strays grew every year. Some dogs lived in subways, begging for scraps; others hunted in packs, scavenging from garbage bins and piles of rubbish. The females that gave birth could expect few of their mongrel puppies to survive to adulthood. Although some Muscovites saw these dogs as pests, many treated them with affection – regularly petting and feeding the same animals. It was these tough, streetwise and friendly dogs that scientists from the Moscow Aviation Institute had in mind for their top-secret space research programme.

The Soviet Union had a long history of using dogs in scientific research. At the beginning of the twentieth century, Ivan Pavlov had carried out famous experiments on the circulation, nervous system and behaviour of canines. Dogs were companionable and easy to train. And, unlike domestic pets, who would miss the odd stray if it disappeared?

So, in the summer of 1950, biologists from the Aviation Institute spent their evenings driving around the streets of Moscow looking for stray dogs. They had a specific list of

A Soviet space dog wearing a pressure suit and acrylic glass bubble mask. Female dogs were chosen due to their size and temperament, but also because of the design of the spacesuits.

requirements. The dogs needed to weigh less than 7 kilos and be small enough to fit into the nose cone of a rocket. They had to be passive, friendly, intelligent and light in colour – or at least with clear patches of white – so they would show up easily on the space-capsule film (and later, on TV) cameras. Females were preferred (all the later dogs were female) as they could urinate without cocking their legs. The scientists concluded that this might be an issue in weightlessness, and there was little room in the suits or cabins.

Whenever the scientists spotted a potential stray they jumped out of their cars with sausage meat in hand, trying to entice the dog into the vehicle. Anyone witnessing the strange scenes would have been wise to keep quiet.

The scientists took the captured dogs to a requisitioned mansion in a Moscow suburb. Here, they cleaned and fed the animals and housed them in individual kennels. More significantly, perhaps, they gave each dog a name – these were no longer anonymous street dogs. Once dozens of animals had been collected, the training programme began.

Fitted with individually hand-sewn flight suits, the small dogs were strapped to vibrating tables and subjected to increasing levels of aircraft noise to simulate launch. They were placed in centrifuges – usually used to train fighter pilots – and spun at up to ten times the force of gravity (10G). They were locked in confined spaces to get them used to space capsules, and low-pressure chambers to see how their bodies would cope with altitude. The compliant dogs were also put through a series of medical procedures. As well as taking X-rays and regular cardiograms, vets operated on the animals to realign the main artery in their necks so it was closer to the skin. This would make it easier for them to monitor the dogs' pulse rates during flight.

Between all this training, the animals led a happy and relatively pampered life. The scientists and vets gave the

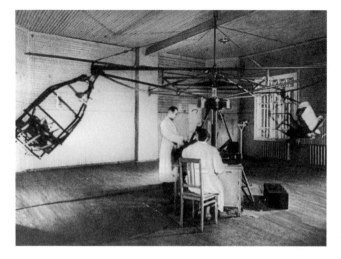

dogs rewards for good behaviour, two meals a day and plenty of exercise. Everyone enjoyed working with the dogs, and many scientists took the animals home at the weekends to meet their families or play with their children.

After months of selection and testing, 14 dogs were chosen for the next stage of the programme. They were shipped out to new kennels at a secret rocket test site in the desert at Kapustin Yar near Stalingrad (now Volgograd).

In the early hours of 22 July 1951, the first two dogs were ready for blast-off. Tsygan ('Gypsy') and Dezik were cute and friendly dogs, probably terrier crosses with maybe a bit of husky. They were strapped onto individual stretchers and loaded, side by side, into a sealed compartment at the top of a cigar-shaped experimental rocket. It was a procedure the dogs had been through many times in training.

Two dogs were flown together so the animals had some companionship. It also gave the scientists more data to work with. Hooked up to sensors and observed by cine cameras using lighting and mirrors, the dogs would be monitored every step of the way.

A Soviet space dog is retrieved unharmed after safely returning to Earth from a sub-orbital flight via a parachute and capsule.

Just before dawn, chief rocket engineer Korolev gave the order to launch the rocket. As it shot away from the pad, Tsygan and Dezik struggled against the pressure of acceleration. They could barely lift their heads, and their breathing and pulse rates increased dramatically. Then, 110 kilometres above the Earth, out of fuel and beyond the atmosphere, the compartment containing the dogs separated from the spacecraft and began to fall back to the ground. During these 220 seconds, the animals would have experienced a feeling of weightlessness before the capsule accelerated towards the Earth. As the dogs neared the ground, parachutes opened to slow their fall.

Tsygan and Dezik had become the highest ever living creatures. But, until the scientists opened the capsule hatch, no one knew if the dogs had survived the adventure.

The team rushed to recover the space capsule. When they opened the hatch, not only were the dogs alive, they seemed to be no worse for wear from their adventure. The handlers released the animals from their straps, hugged them, and fed them water and sausage meat. Tsygan and Dezik were overjoyed – quite a different reaction to the American chimps.

A few days later, on 29 July, Dezik was once again strapped into the top of a rocket for a second of these sub-orbital parabolic flights. The plan was to test the same dog to see if repeated spaceflight caused any long-term effects. Dezik was accompanied in the capsule by another dog, Lisa ('Fox'). Korolev gave the command to launch and, once again, the rocket shot into the sky high above the desert. But, as the dogs' compartment plummeted towards the ground, no parachute opened. The animals were killed. The team was devastated and Dezik's companion from the first flight, Tsygan, was retired from the programme and lived out his life for another ten years at the home of space scientist and diplomat Anatoli Blagonravov.

Just over a month later, a soldier was taking one of the dogs for a walk when it ran away. In a desperate attempt to fill her place in the capsule, and fearing the wrath of Korolev, the scientists spotted a nearby stray hanging around outside the canteen. Enticed with the inevitable sausage meat, the rocket team fitted the remarkably passive and friendly dog with sensors, strapped it into the capsule and blasted it into space. ZIB – or Substitute for Missing Bobik – made it back to Earth alive.

Under constant pressure from the Kremlin to build missiles to carry nuclear warheads, rather than people, Korolev nevertheless persisted with the experimental rocket programme. He eventually flew almost 30 of these sub-orbital missions with dogs on board but, as the rockets got bigger and more powerful, the chief engineer had something much more ambitious in mind.

Йка

Laika

This trusting little stray could never have anticipated that she would one day float 200 miles above the Moscow streets. Nicknamed Laika, meaning 'barker', she would be canonised as a proletarian hero, her place in history forever assured.

Laika: the First Space Hero

3 November 1957, Baikonur Cosmodrome

By 1957 the Soviet missile and space programme had relocated to a larger and even more isolated site in the Kazakh steppe, which was better suited to the giant rockets Korolev was now developing. Whereas the sub-orbital launchers carrying the first dogs into space could only reach speeds of some 3 km/second before falling back to Earth, the new rocket would achieve 11 km/second – enough to escape the planet's gravity and reach orbit.

On 4 October Korolev launched the first satellite, Sputnik 1. Circling the Earth every 90 minutes, the incessantly beeping silver sphere sent America into a frenzy. The fact that the Soviets could place such a sizeable object in orbit meant they were more than capable of delivering a nuclear missile to anywhere in the world.

As the US struggled to get even a pineapple-sized satellite off the ground (only finally managing a successful launch in January 1958), the Soviet leadership was taken aback by the worldwide impact and propaganda coup of Sputnik 1. Soviet Premier Nikita Khrushchev demanded another spectacle within a month, just ahead of the fortieth anniversary of the Bolshevik revolution.

Laika in the training capsule prior to her space mission. As part of their preparation for flight, dogs were trained to be comfortable in confined spaces and endure extended periods of inactivity.

Fortunately, Korolev had something up his sleeve. Alongside the technologically primitive radio transmitter of Sputnik 1, Korolev had been quietly developing a far more advanced spacecraft: a capsule for a space dog.

Sputnik 2 weighed more than half a tonne – yet another sobering statistic for the US in terms of warhead capabilities. At the top, the new padded and windowless space dog compartment was fitted with air-conditioning systems in order to maintain breathable air and constant temperature. Although it was cramped, there was enough room for a dog to stretch its legs.

For the first time, the spacecraft was also equipped with a feeding system. This delivered a nutritious gel made up of ground meat and water. A vacuum cup device was fitted to the animal's backside to remove waste. As before, the dog would be covered with sensors so its vital signs – and even movement – could be transmitted back to Earth.

But, for all the complexity of this new space capsule, there was a fundamental problem. Although Korolev had

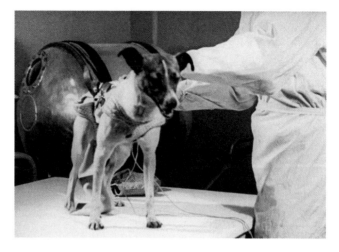

the rockets to launch large objects into orbit, he didn't have the technology to get them safely back to Earth. The dog making the journey into space would therefore provide valuable data on the effects of weightlessness, but it wouldn't be returning.

Like her predecessors, Laika had been plucked from the streets of Moscow. Of uncertain heritage – likely part husky, part terrier – but affectionate, patient and with dark expressive eyes, Laika was trained to become accustomed to living in her spacecraft. She had learned to trust her handlers, and before she was sealed into the top of the rocket, one of them kissed her nose to wish her goodbye.

On 3 November, Laika blasted off from the new Baikonur Cosmodrome and accelerated into orbit. Readings of her pulse and breathing suggest she was frightened by the initial acceleration but relaxed when the capsule reached orbit and she became weightless.

News of the record-breaking space dog made the front pages around the world, with pictures of the happy hound (taken before the mission) supplied by the Soviet news agency TASS. Many articles published outside the Soviet Union included speculation about whether the dog would ever return to Earth. Only those closest to the space programme knew the truth.

It appears the original plan was to allow Laika to orbit the Earth for seven days and then feed her poison, so she would die painlessly. However, it's unlikely poisoned food was ever included in the mission, so maybe the engineers intended that the dog died when the oxygen ran out. Either way, she suffered a far more tragic death.

Within minutes of the spacecraft reaching orbit, the living compartment began to overheat. The cooling fans couldn't cope with the extreme heat of the unshielded Sun. With temperatures eventually reaching more than 40 degrees Celsius, it is now certain that Laika died after about two hours, from panic and heat exhaustion.

At the time, no one in Russia or elsewhere – except those working directly on the mission – knew the truth. And, for their own safety, those involved weren't going to admit it to anyone. In fact, if the rocket had blown up during launch, the whole mission would have been hushed up and another dog flown.

For days after the rocket reached orbit, Soviet media declared that Laika was alive and well. They even claimed that she was calm and enjoying the mission. By that time, the dog was long dead. Five months later, her remains were destroyed when the satellite burned up as it re-entered the Earth's atmosphere.

In the US, animal rights groups protested the flight and the inherent cruelty of the one-way mission. But across the Soviet Union and Eastern Europe, Laika was celebrated as a heroine. And rightly so. Although she didn't choose her fate, she had proved that animals could live in the weightless conditions of outer space.

A 1950s shop window display of Laika cigarettes. When American U2 pilot Francis Gary Powers was shot down over Sverdlovsk in May 1960, he was reportedly offered a Laika cigarette by a villager.

Laika was the first of the space dogs to be commemorated in books, on stamps and postcards. You could buy Laika-branded cigarettes and matches, and her image was printed on ornaments, paperweights, clocks and watches. To tie in with Soviet anniversary celebrations, Laika was often portrayed looking noble next to Lenin or other heroes of the revolution (or at least those who hadn't been airbrushed from history). The dog was presented as though she had known she was dying for an important cause: helping her masters conquer outer space.

But, although the sub-orbital flights continued, orbital flights were put on hold until Korolev could perfect a means of bringing animals back alive.

At a Moscow press conference in 1998, Oleg Gazenko, a senior Soviet scientist involved in the project, lamented, 'The more time passes, the more I'm sorry about it. We did not learn enough from the mission to justify the death of the dog …'. The spacecraft had not been designed to be recovered and it burned up on re-entry on 14 April 1958.

Desktop souvenir trophy fashioned to resemble Laika's orbiting space capsule, Sputnik 2.

One of a series of desk clocks that featured space heroes including Laika, fellow space dogs Belka, Strelka and Zvezdochka, and cosmonauts Vladimir Aleksandrovich Shatolov and Vladimir Mikhaylovich Komarov.

Wood and acrylic desktop souvenir clock featuring Sputnik 2 and Communist red stars. This would have been presented to a worker, the inscription reading 'To Aleksandr Ivanovich Churin on his 60th anniversary, from the factory'.

"Лайка была замечательно собакой"

"Laika was a wonderful dog ... quiet and very placid. Before the flight to the cosmodrome I once brought her home and showed her to the children. They played with her. I wanted to do something nice for the dog."

— Dr. Vladimir I. Yazdovsky, Soviet Space Programme

Blown-glass commemorative ornament
showing Laika at the feet of a soaring
Sputnik 2.

Desk calendar from 1957 with rotary
dial for changing the days and months.
The front features an embossed Laika
and a model space capsule perched on
top. Text on the reverse urges workers of
the world to unite.

One of the most popular canvases for Laika's image was the cigarette case. These ranged from inexpensive tin boxes to more finessed plated metal and leather cases. Many featured Laika alongside Communist emblems and portraits of Lenin.

BELOW Vintage Riga cigarette lighter decorated with Laika in bas relief.
OPPOSITE Laika cigarettes produced under supervision of the Ministry of Food Industry. Introduced in 1957, the brand was finally discontinued in the 1990s.

BELOW Wooden trinket boxes. The top box has an inscription to a Russian worker from the Lenin Komsomol car plant in Moscow. The bottom box features the Soviet war memorial statue in Berlin's Treptower Park.
OPPOSITE Halva confectionery tin from the Marat Moscow sweet factory.

BELOW 1950s commemorative acrylic and metal desk/table star in a presentation box. The initials 'AMKOC' stand for 'Association of Museums of Cosmonautics'. **OPPOSITE** Pen desk set with slot for holding letters. **OVER THE PAGE** Space dog being fitted with a flight suit.

"Чем больше времени проходит, тем больше я сожалею об этом"

"Work with animals is a source of suffering to all of us. We treat them like babies who cannot speak. **The more time passes, the more I'm sorry about it.** We did not learn enough from the mission to justify the death of the dog."

— Dr. Oleg Gazenko, Director of the Institute of Biological Problems (speaking in 1998)

BELOW 1958 postcard with portrait of Laika by the artist E. Gundobin, with the first three sputniks shown in the background. **OPPOSITE TOP** Japanese collectable playing card and Soviet matchbox label. The matchbox label reads 'The first passenger of Sputnik, the dog Laika.' **OPPOSITE BOTTOM** Soviet sweet wrappers.

OPPOSITE Model depicting Laika in the specially designed space capsule inside Sputnik 2. **BELOW** Silver charm for a bracelet or pendant (actual size is a tiny 1.5 x 2cm [⅓ x ¾ in]).

Actual size

ка и
елка

Belka & Strelka

These plucky pups would become the first Soviet superstars. In a regime that eschewed celebrating individual achievement, they were hailed as heroes, with press and merchandise that would be the envy of present-day pop stars, and books and films in their honour.

Celebrity Dogs

19 August 1960, Baikonur Cosmodrome

By the fourth orbit, it looked like the mission had failed. Belka and Strelka had launched alongside two rats, 40 mice and a selection of insects, cultures of microbes and seeds. As Korolev's powerful new rocket – a prototype for the first manned spaceflight – blasted off from Baikonur, the dogs' medical data appeared to be normal. Or at least as normal as could be expected. But the live TV pictures beamed back from the Sputnik 5 capsule told another story: the dogs were not showing any signs of movement. Finally, after about five hours in space, Belka started vomiting and the dogs began to wake up.

Sputnik 5 was the second attempt to send dogs into orbit and return them safely to Earth. Belka ('Squirrel') and Strelka ('Little Arrow') had been specially selected during another recruitment exercise on the streets of Moscow. Only this time, the requirements for the mission were more stringent than before. As well as being female, compliant, friendly and brightly coloured for the onboard TV cameras, the canine candidates could not be larger than 34 cm (13 in) high and 43 cm (17 in) long in order to fit in the cramped orbital compartment.

The biologist tasked with selecting the dogs, Dr. Ludmilla Radkevich, armed herself with sausage meat and a tape measure and set off with a driver to find suitable strays. Whenever they spotted a likely dog, Dr. Radkevich would

Dr. Oleg Gazenko, the director of Russia's Institute of Biomedical Problems, victoriously holds Belka and Strelka aloft during their first press conference.

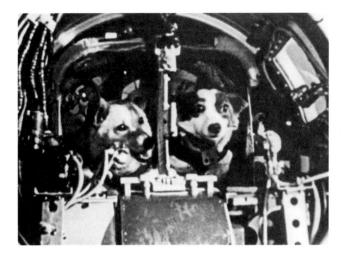

leap out of the car to check its dimensions. Eventually 60 new space dogs were inducted into the training programme.

The spacecraft designed for the dogs was far more sophisticated than Laika's. As well as TV cameras, it incorporated solar cells for power and a refrigerated cooling system to dissipate the Sun's heat. As before, the animals would be fed from an automatic feeding device with a gelatinous mixture of food and water, and their waste collected in sealed pouches. Crucially, unlike Laika's mission, the plan was to return the dogs to Earth unharmed.

The first dogs selected to fly were Bars ('Panther') and Lisichka ('Little Fox'). Korolev visited the animals every day to check on their welfare and had taken a shine to ginger-coloured Little Fox. As the dogs were sealed into their capsule on 28 July, he told her he very much hoped she'd return. A few seconds after launch, the rocket exploded, killing both of the dogs. Korolev was visibly upset but, unlike with US mission failures, there was no publicity.

Now, less than a month later, Belka and Strelka were weightless in orbit. The medical data looked good and the

After a 24-hour orbit, 200 miles above the Earth, the animals returned to the ground safely. The little dogs became instant celebrities, and were touted before the world's media in a Soviet propaganda coup.

TV cameras showed them moving around, barking and seemingly perfectly content. The question now for Korolev was: could he get them back?

After 18 orbits, he gave the command to fire the retro rockets to slow the spacecraft down and drop it out of orbit. Belka and Strelka plummeted to Earth at more than 10G. They jarred from side to side as the cannonball-shaped capsule was buffeted violently by the atmosphere.

At 7,000 metres, their compartment ejected from the capsule and the parachutes opened. They floated to the ground within 10 kilometres of the intended landing site – a remarkable feat of precision. The dogs were completely unharmed and wagged their tails to greet their masters.

Thanks to Belka and Strelka, Korolev now knew that animals could be launched into space, survive in orbit for more than a day, and return safely to Earth. He now planned to attempt the same thing with a human, if the Americans didn't beat him to it.

Belka and Strelka's flight was another propaganda coup for the Soviet Union and, after their safe recovery, pictures

of the dogs living, seemingly happily, in outer space were broadcast on state television. The public reaction was overwhelming – here were two of the cutest and happiest dogs, back on Earth after flying in space. Once again, Soviet space dogs were leading the way to the stars.

As the victory car carrying Belka, Strelka and biologist Dr. Radkevich made its way through Moscow, people in neighbouring vehicles and along the roadside applauded and congratulated them. Whenever they stopped, crowds gathered to catch a glimpse of the animals. The dogs were introduced to VIPs and appeared with their handlers on TV as chat show guests. Everyone wanted to be seen with them.

Laika souvenirs had been popular, but Belka and Strelka took the commemorative space dog business to a whole new level. In a land that officially shunned the trappings of glamour and celebrity, the dogs were superstars with all the mass-produced memorabilia to match. And whereas the earlier Laika souvenirs had had a certain dignity to them, these were of a more playful and cartoonish nature.

The dogs were portrayed smiling in spacesuits, peering out of rocket portholes or driving side by side in stylised space ships, whizzing around among the stars. There were figurines, animations, books and stories. No one ever knew about Bars and Lisichka, or the subsequent deaths, in the months leading up to Yuri Gagarin's flight, of space dogs Pchyolka ('Little Bee') and Mushka ('Little Fly').

Wall clocks with various display
faces featuring pressure and
temperature gauges.

BELOW 1960s wood-mounted clock/barometer, one of a series featuring space dogs and Soviet cosmonauts. **OPPOSITE** Wooden desk clock, one of another series featuring both canine and human cosmonauts.

BELOW 1960s alarm clocks. **OPPOSITE**
Molnija desk clock engraved with the
date of the dogs' voyage and depicting
Sputnik 5 soaring into space.

The inscription reads
'USSR "East–1" 12.04.1961'.

Mechanical wristwatch from the 1960s with decorative bezel. Belka and Strelka featured on hundreds of wristwatches by famous manufacturers, including Molnija, Poljot, Pobeda and Vostok (official supplier to the Soviet military).

Souvenir coin (enlarged to show detail) from the Memorial Museum of Cosmonautics in Moscow. One side shows Sputnik 5 taking off, and the text reads 'Museum of Cosmonautics'; the other side depicts Belka and Strelka in their classic skyward-gazing pose, bearing the legend, 'Dogs – Cosmonauts'.

Belyanka and Pyostraya in the Rocket,
a 1961 Russian children's book by writer
V. Borozdin, describing the work of the
Soviet space dogs.

'The cockpit looked
identical to the one
they trained in every
day. The dogs settled
in straight away as it
felt familiar, and they
calmly lay down. How
could they suspect that
they were being placed
in a rocket that was
going to take off in a
few minutes and carry
them into outer space?'

Из. првых. особой тренировках."

— Шарлатанов жгением, да и нарвём. (съели)

снова не легла. Прошло ещё несколько дней, и Белянка
усвоила правило: чтобы получить кусочек сахару, нужно влезть в кабину и лечь. Когда же сахар подъедет
к самому носу, чтобы съесть его, ей нужно вскакивать.
А если вскочишь, то нового уже не получишь.

Так шаг за шагом Белянка познавала то, что от неё
требовали. А требовали от неё каждый день всё больше
и больше. Едва она успевала усвоить одно, как ей пред-
лагалось другое.

Как-то Наташа, которая часто осматривала Белянку и других собачек, прицепила ей на шею и к лапам
какие-то коробочки. Белянка не знала, что это были

пать, толкать её, недовольно повизгивая. Ничего не
помогало. Дверца не открывалась. И лента всё ещё не
двигалась.

Так продолжалось очень долго, пока Белянка не
устала и не легла. И — странная вещь, — как только
она легла, лента снова передвинулась, и на ней был
сахар. Белянка вскочила, съела его и стала ждать. Но
лента опять не двигалась — до тех пор, пока Белянка

12

Из. людских поз. леза! Слопать их собачки... (Белянка и Пёстрая в кабине (поезд).

конечного сооружения. Кабина была точно такая же,
как и та, в которой они каждый день проводили трени-
ровки. Собачки почувствовали себя в ней как дома и,
как полагалось, сразу же преспокойно улеглись. Они
и не подозревали, что их поместили в ракету, которая
через несколько минут должна взлететь на огромную

22

высоту. К шее и лапам, как и во время тренировок, им
прицепили коробочки, и дверца кабины захлопнулась.

Они уже не слышали, как снаружи отдавались по-
следние приказания, не видели, как, осмотрев ракету
и проверив, всё ли в порядке, люди быстро скрылись.
в специальном бетонированном укрытии.

Положив голову на лапы, Белянка и Пёстрая спокойно ждали, когда же к их носу подъедет кусочек
сахару.

И вдруг всё взревело. Кабину рвануло, подбросило
и понесло... Белянке захотелось выскочить из кабины,
как когда-то в начале тренировок. Но где там! Неведо-

Нестойко сам клещи Белянку и Пёструю в носу. Это ракета двигалась. вверх.

'Suddenly everything
was loudly roaring. The
cockpit was shaking and
moving at speed. Belka
suddenly felt the urge to
jump out of the cockpit
like she used to do when
she was training, but it
wasn't possible!'

Set of five handmade and hand-painted wooden nesting Matryoshka dolls. The largest is 15 cm (6 in) high.

"Белка лидер в команде"

"Belka is the leader in the team, the most active and sociable. In training she showed the best results. Strelka — a mongrel of a light colour with brown spots — is timid and slightly reserved, but, nevertheless, friendly."

— Mitrofan Ivanovich Nedelin, Chief of
the Strategic Missile Forces

Porcelain commemorative plate showing Belka and Strelka looking skywards, with Saturn and Sputnik 5 picked out in gold.

Hand-painted Gzhel porcelain figurines
of Belka and Strelka wearing cosmonaut
suits and breathing apparatus.

OPPOSITE Belka and Strelka at their first press conference. **BELOW** Porcelain figurines from the Dmitrovsky Porcelain Factory (Verbilki), 1960. They are wearing their distinctive green and red jerseys.

Night lamps from the 1960s,
featuring Dimitrovsky porcelain
figurines and toy cosmonaut figures.

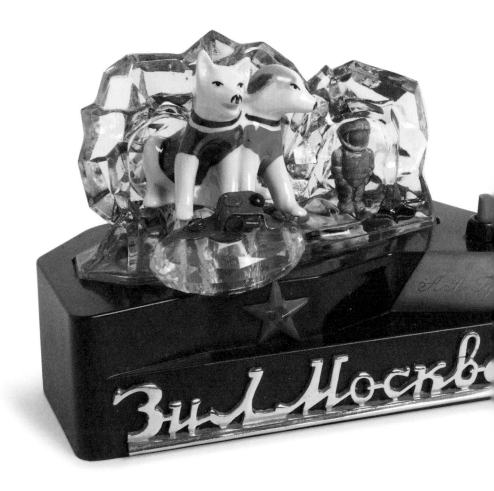

"коллеги и друзья"

"Any experimenter working with animals does not see them as dogs. He sees them as his **colleagues and friends.** And the amazing thing is, those little nicks or fur cutting for the introduction of sensors were never perceived by the animals as an aggressive, unfriendly act. On the contrary, sometimes they turned and licked our cheeks."

— Dr. Oleg Gazenko, Director of Moscow's Institute
of Biomedical Problems

ОГОНЁК № 35 АВГУСТ 1960
ИЗДАТЕЛЬСТВО «ПРАВДА»

КОСМОС! ЖДИ В ГОСТИ СОВЕТСКОГО ЧЕЛОВЕКА!

Одиннадцать картин И. И. Левитана

Почему плакал маленький американец?

Cover of the Soviet magazine огонёк
('Little Light') from August 1960,
showing the dogs gazing heroically
starwards. The headline declares
'Space! Waiting for Soviet man to visit!'

Magazine spread from огонёк. The headline reads 'Intergalactic Travellers'.

ОГОНЁК

Фото Ю. Кривоносова

А. ГОЛИКОВ, Н. СМИРНОВ

Специальные корреспонденты «Огонька»

Вот они — первые межзвездные путешественники, благополучно возвратившиеся на Землю, — говорит врач-экспериментатор.

[column of body text, largely illegible]

Их голоса зазвучат в эфире. Белка и Стрелка «дают интервью» для радио.

МЕЖЗВЕЗД

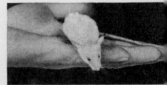

Когда на собак надевали «полётную одежду», они себя вели спокойно. Но они стали повизгивать, лизать руки при посадке на мороз.

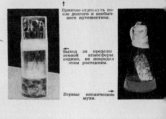

Приятно отдохнуть после долгого и необычного путешествия.

Выход за пределы земной атмосферы, очевидно, не повредил этим растениям.

Первые космические мухи.

The photo captions read 'It's nice to finally relax after a long and unusual trip', and 'Travel in space didn't harm these plants'.

Captions read (left to right) 'Brave – be disciplined, listen to the onboard equipment!'; 'Careful! There are dogs there!'

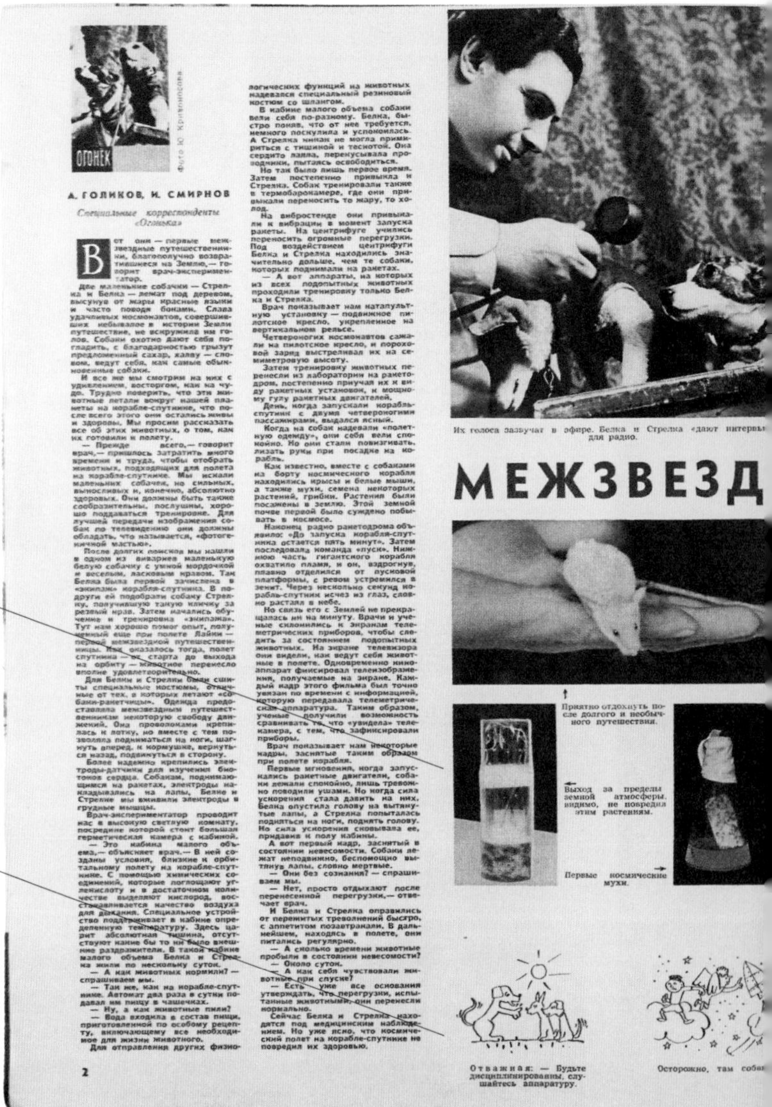

Отважная: — Будьте дисциплинированы, слушайтесь аппаратуру.

Осторожно, там собаки.

2

Photograph showing the dogs meeting the public ahead of their press conference in Moscow.

Сердцебиение Земли

Мечту
стрелою заострили,
в цехах
в броню
отлили мысль,
Из молний
выковали крылья,—
и вот взвилась жар-птица ввысь.

Как песнь,
как искра звезд московских,
что пульсирует вдали,
я ловит лунным ухом
космос
сердцебиение Земли!

Иван АНДЕНКО

Фото Ю. Кривоносова,
Ю. Абрамочкина
и В. Николаенко.

Самые знаменитые собаки на земном шаре.

Е ПУТЕШЕСТВЕННИКИ

В час рассвета

озвездия блуждали
в вечной мгле,
азалось, им безмерный
счет потерян...
Мы первые сумели
на Земле
Открыть вселенной
запертые двери.

За облаками в светлом
серебре
оржественно приходит
час рассвета...
еперь я знаю — на любой
заре
аемка знамени моей
Страны Советов.

Михаил СВЕТЛОВ

Белка: — Мне в космосе понравилось.

A cartoon depicting Belka and Strelka with a caption that says, 'This is the way I train daily!'

— Вот так я тренирую-
юсь ежедневно.
Рис. Б. Семенова.

Captions read (left to right) 'How do you feel Strelka? Much lighter than on earth!'; 'Consider this: there could be cats among the ones meeting us!'; 'Come back home with victory'; 'Warm welcome'.

— Как себя чувству-
шь, Стрелка?
— Легче, чем на Земле!

— Учти, нас могут
встречать и кошки!

Возвращение с победой.

Теплая встреча.
Рисунки В. Сигачева.

Large keepsake/storage box decorated
with depictions of Laika, Belka and Strelka,
and Lenin, and bearing the date of the
Sputnik 5 mission.

The inscription reads, 'For the Chairman of the Central Committee of the Trade Union of Workers of the Aviation Defence Industry, Comrade Kareev A.T'

BELOW Large confectionery box from
the Babayevsky chocolate factory.
OPPOSITE Small confectionery tins.

Cover of the 1961 children's
book *The Adventures of Belka
and Strelka*, by Yuri Galperin,
published by Children's World.

'In a little cage there are other passengers – white mice, as usual nibbling on rusks. In another cage, there is a big rat drinking water, and in a glass jar, flies. Next to them, under glass, is a green plant. This is how many passengers there are on board the space rocket!'

Ракета летит всё вперёд и вперёд. Теперь это уже маленькая звёздочка на небе, а телевизор показывает всё, что делается в кабине. Вот и собаки привыкли к полёту. Они спокойно едят желе.

36

Рядом с ними в маленькой клетке другие пассажиры — белые мышки. Сидят и, как ни в чём не бывало, грызут сухари. Пьёт воду в своей клетке большая крыса, в стеклянных баночках — плодовые мушки, а рядом с ними, под стеклом, зелёный цветок. Вот сколько пассажиров в ракете!

37

'The space rocket is flying further and further away. Now you can only spot it as a small star in the sky, but on the screen you can see what is happening inside the cockpit. The dogs finally settle on the flight and calmly eat jelly.'

'The flight is over! They landed safely. The cockpit is open. Healthy and happy space travellers can see their favourite doctor again! As usual, he has a treat for them, this time Halva! Now Belka and Strelka are the most famous dogs in the world, they are the first dogs to complete such a flight.'

Полёт окончен! Кабина открыта. Здоровые и счастливые, путешественницы снова видят любимого доктора. Как всегда, у него приготовлено для них лакомство — на этот раз халва.

Теперь Белка и Стрелка — самые знаменитые собаки в мире. Они первыми совершили такой полёт.

40

Во всех газетах мира на первых страницах портреты отважных космонавтов. «Неужели это всё про нас? — думают Белка и Стрелка. — Как жаль, что мы не умеем читать, хотя бы по складам, как первоклассники...».

Знаменитых космонавтов приглашают в радиостудию «выступить». Но, увидев раз-

41

Faux tortoiseshell cigarette cases
featuring Belka alongside St Petersberg's
bronze statue of St Peter the Great (left)
and Strelka next to the Kremlin.

Soviet book *Ship Back from Space*, by
Boris Valerianovich Lyapunov, 1960.

'During the flight, the dogs were completely cut off from the outside world. However, they could see each other through transparent mesh. Belka and Strelka have been friends for a long time, and each was not deprived of the pleasure of looking at their travel companion'.

'When the camera was turned on, it showed Belka and Strelka becoming anxious, but they settled down straight away'.

'The satellite ship was spinning around the Earth. It made the round-the-world trip in an hour and a half! If you could only have seen your home planet from there, you wouldn't have recognised it!'

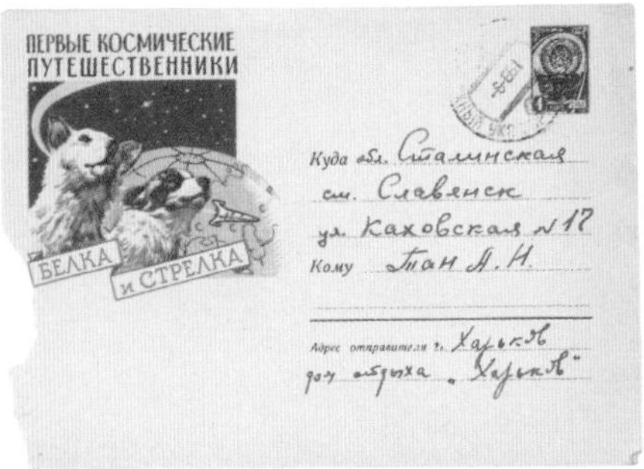

top 1960s New Year decorations featuring Belka and Strelka alongside seasonal motifs. These would be strung as a garland.

above Postal cover envelopes from 1961; the illustration on the left is by the artist Zavjyalov.

opposite Postcard showing the dogs at their first press conference; postcard by the artist L. Aristov from 1972.

Белка и Стрелка

Space Dog Diplomacy

June 1961, Vienna

Two months after Gagarin's triumphant return to Earth, Khrushchev had a lot to boast about and took every opportunity to ridicule America's space efforts. It was clear that in the early 1960s – and largely thanks to the sacrifices of space dogs – the Soviet Union was ahead of the US.

On 5 May 1961, America had finally managed to get an astronaut, Alan Shepard, off the ground and back to Earth but it was only on a 14-minute sub-orbital flight – not unlike Korolev's early space dog experiments. It wouldn't be until the following February that US astronaut John Glenn would make it into orbit.

The Vienna Conference of 1961 was the first opportunity for Khrushchev to meet America's recently elected president, John F Kennedy. With both countries ratcheting up their nuclear arsenals, there was much at stake. Accounts of the conference suggest the summit didn't go well. Both leaders were wary of each other, but during dinner, Jackie Kennedy turned to the subject of space dogs.

By now Strelka had given birth to a healthy litter of puppies. The happy event had been shared in newsreels shown across the Soviet empire. As well as providing more good publicity for the space programme, Korolev was keen to study the puppies to see whether the space environment, particularly radiation, led to any hereditary genetic effects. Jackie Kennedy suggested that Khrushchev send her one of

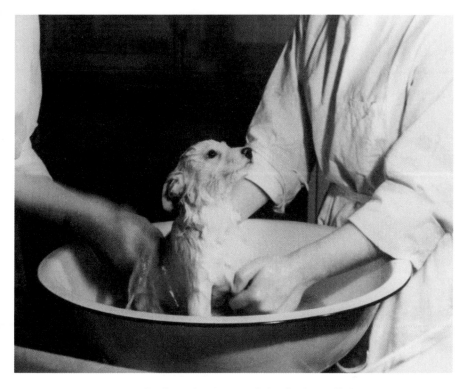

Kudryashka, one of Strelka's litter of six 'pupniks' by Pushok, who was also in the space dog programme but only took part in ground-based experiments.

FOLLOWING PAGES Strelka with her litter.

the dogs – inadvertently beginning a diplomatic process that may have helped prevent a third world war.

A few weeks later, the White House received an unexpected delivery from the Soviet embassy – a puppy with a Russian passport. Fearing this could be a covert spying operation, the FBI combed through the fur of the tiny dog, checking it for surveillance bugs. But space puppy Pushinka – meaning 'fluffy' – was allowed to stay, joining the young First Family's menagerie of other pets. Unfortunately though, the President himself had to avoid the dog as it made him sneeze and come out in a rash.

Pushinka was one of several gifts and letters the two leaders exchanged in a series of back-channel communi-

ABOVE Strelka's puppy Pushinka posing with her Soviet passport upon arrival at her new home, the White House.

LEFT Kennedy family dog Charlie with Pushinka on the White House lawn in 1961.

cations, which helped foster warmer relationships between them. In a letter to Krushchev in June 1961, Kennedy wrote: 'Mrs Kennedy and I were pleased to receive "Pushinka". Her flight from the Soviet Union to the United States was not as dramatic as the flight of her mother, nevertheless, it was a long voyage and she stood it well'. When the Cuban Missile Crisis threatened to kick-off World War III in October 1962, having Pushinka in the White House may have played a role in helping the leaders step back from the brink.

Pushinka also made her own more direct contribution to thawing East–West relations. White House Welsh terrier Charlie took a particular shine to the Russian female and, after a short Cold-War romance, in June 1963 the pair became the proud parents of a litter of puppies. Or, as the President referred to them, 'pupniks'.

Rather than be over-run by dogs, the Kennedys decided to give two of the puppies away to American children. More than 5,000 kids wrote letters giving reasons why they should home the pupniks. 'I'm so happy,' ten-year-old Karen House of Westchester, Illinois, told reporters in August when she learned she was to receive female Butterfly, one of the space dog descendants, 'I'm all mixed up!'. Meanwhile, male puppy Streaker went to nine-year-old Mark Bruce from Columbus, Missouri, who said he 'had prayed for another dog' after his first died in an accident.

When President Kennedy was assassinated in November 1963, Pushinka was passed on to a White House gardener and she mothered another litter of puppies. It's likely that somewhere in America, descendants of those dogs are alive today. Their owners may not even be aware of their incredible space heritage.

еские
ческих
бак

Legacy of the Space Dogs

The contribution of dogs to the space programme cannot be overstated. Even after the first successful human flight, and to this day, they continue to be celebrated in Russia and throughout the world.

Moon Dogs

March 1966, outer space

When Yuri Gagarin parachuted to Earth in April 1961 – using the technology pioneered by Belka and Strelka – he succeeded them as a new national space hero. There was no longer any need for the corps of space dogs. The programme was gradually wound down as Korolev's team notched up new achievements in human spaceflight.

But, as the superpowers raced to be the first to the Moon, the Russians planned one last canine space effort. A mission that would boldly take dogs further than they had ever gone before.

On 22 February 1966, space dogs Ugolek and Veterok had launched into orbit in an adapted Voskhod capsule – a larger and more sophisticated spacecraft than the one that had carried Gagarin. After three weeks, the dogs were still in space, alive and seemingly well.

The mission took Ugolek and Veterok high above the Earth, through the Van Allen radiation belts, and was designed to test the long-term effects of spaceflight on the body and impact of potentially damaging space radiation. The dogs were fitted with a new generation of sensors – including devices surgically implanted in their hearts – and were automatically fed and given medication.

In their windowless capsule, adrift in the cosmos, Ugolek and Veterok might have wondered if they would ever again know a life where they could run beneath the sky or

Strelka, Dymka ('Smoky') and Chernushka ('Blackie') at a TASS press conference, 1961.

smell anything other than each other and the stale air they breathed. Eventually, after 22 days, the retros fired and the dogs came back to Earth. The dogs' first steps were shaky but they survived physically unscathed. Newsreel footage captured after the flight shows them looking subdued and suggests they were still traumatised by the experience.

Ugolek ('Little Piece of Coal') and Veterok ('Light Breeze') being exercised in their pressurised flight suits. They would go on to spend 22 days orbiting Earth aboard the Kosmos 110 biosatellite.

The legacy of Ugolek and Veterok's mission, however, was a better understanding of long-duration spaceflight. Ultimately, the dogs helped enable the Soviets to establish space stations and, with the results of the experiment published around the world, added to NASA confidence in sending astronauts through the Van Allen belts to the Moon.

Ugolek, Veterok, Laika, Belka and Strelka paved the way for the first woman in space, Valentina Tereshkova, the first spacewalk, and the first mission with a crew of three. They helped man reach the Moon and establish orbiting settlements beyond the Earth. The dogs had no choice in their endeavour but were affectionately cared for by the scientists and engineers pushing at the final frontier.

The legacy of the Soviet space dogs lives on in the souvenirs and memorabilia of the early space race. In 2008, a monument to Laika was unveiled in Moscow, and some of the dogs' space capsules are on display in the city's Cosmonautics museum. More importantly, today's astronauts living for months on end on the International Space Station, or those training for a return to the Moon and Mars, still owe their comfort and safety to the pioneering efforts of stray dogs.

As we move beyond the Earth to eventually colonise other planets, maybe future spacefarers will choose to take dogs with them. Man's best friend has been with us since the dawn of civilisation. Given their significant contribution to spaceflight already, it would be only fitting that dogs accompany us on this next great adventure.

OPPOSITE Soviet space dog Otvazhnaya ('Brave One') with rabbit Marfusha ('Little Martha') after their sub-orbital flight in 1959. **BELOW** Desktop commemorative trophy depicting Otvazhnaya, Marfusha and Belka.

BELOW Desktop display model featuring Chernushka ('Blackie'), whose orbital flight aboard Sputnik 9 was made with cosmonaut mannequin 'Ivan Ivanovich'. **OPPOSITE** Alarm clock and desk clock showing Zvezdochka ('Little Star'), who flew on Sputnik 10, also with the mannequin Ivan Ivanovich.

BELOW Confectionery tin featuring
Zvezdochka. **OPPOSITE** Cigarette cases
commemorating Ugolek ('Little Piece
of Coal') and Veterok ('Light Breeze'),
Zvezdochka and Chernushka.

TOP TO BOTTOM Matchbox labels featuring Otvazhnaya and Marfusha; 1960s postal stamps from Poland and Russia depicting Laika, Ugolek and Veterok, Zvezdochka, and Belka and Strelka. **OPPOSITE** Confectionery tin featuring Chernushka.

Desk clocks commemorating Vladimir Aleksandrovich Shatalov, who flew three missions of the Soyuz programme between 1969 and 1971, and Vladimir Mikhaylovich Komarov, who commanded Voskhod 1 in 1964.

Picture Credits

The publisher would like to thank the following organisations for permission to reproduce images in this book. In all cases, every effort has been made to credit the copyright holders and contributors, but should there be any omissions or errors the publisher would be pleased to insert the appropriate acknowledgement in any subsequent edition of the book.

k/barometer featuring the joint USSR Apollo Soyuz crew, the international spaceflight in 1975. to right: astronauts Thomas ford, Donald Slayton and Vance nd and cosmonauts Aleksey nov and Valeri Kubasov.

Clock/barometer celebrating Valentina Tereshkova, Soviet cosmonaut, engineer and politician, and the first woman in space, having piloted the Vostok 6 on 16 June 1963.

Desktop display model of Yuri
Gagarin's space rocket Vostok 1,
with a mounted clock depicting
the cosmonaut's space walk.

"первый человек в космосе или последняя собака?"

"Am I the **first human in space, or the last dog?**" The reported words of Yuri Gagarin upon completing his own space mission.

Yuri Gagarin, Soviet cosmonaut and first man in space